MARION MINING & DREDGING MACHINERY
PHOTO ARCHIVE

Edited by the
Historical Construction Equipment Association

D1452255

Iconografix
Photo Archive Series

Iconografix
PO Box 446
Hudson, Wisconsin 54016 USA

Library of Congress Card Number: 2002112568

ISBN 1-58388-088-7

03 04 05 06 07 08 09 5 4 3 2 1

Printed in China

Cover and book design by Shawn Glidden

Copyediting by Suzie Helberg

COVER PHOTO: See Page 29

BOOK PROPOSALS

Iconografix is a publishing company specializing in books for transportation enthusiasts. We publish in a number of different areas, including Automobiles, Auto Racing, Buses, Construction Equipment, Emergency Equipment, Farming Equipment, Railroads & Trucks. The Iconografix imprint is constantly growing and expanding into new subject areas.

Authors, editors, and knowledgeable enthusiasts in the field of transportation history are invited to contact the Editorial Department at Iconografix, Inc., PO Box 446, Hudson, WI 54016.

About the Historical Construction Equipment Association

The Historical Construction Equipment Association (HCEA), a registered, non-profit organization founded in 1986, is dedicated to preserving the history of construction, surface mining, and dredging equipment. This includes corporate archives, histories of product development and use, descriptive literature, memorabilia, and, of course, the machines themselves. Members hailing from all corners of the world are bound together by their interest in historic construction equipment and a concern for the preservation of these machines and their histories.

One of the most important activities of the HCEA is the publication of a quarterly magazine, *Equipment Echoes*. Featuring a color cover, this magazine contains historical, educational and technical articles, numerous beautiful photographs, and reports on historical equipment news and events.

The HCEA also sponsors annual conventions at various locations in North America. These conventions include displays and demonstrations of historical equipment; historical presentations and movies, toy and memorabilia displays and dealers; and plenty of friendship with people who share a common love for old machinery.

The HCEA has established the world's first public archive and museum, The National Construction Equipment Museum, dedicated exclusively to the history of the construction, dredging, and surface mining industries. Located near Bowling Green, Ohio, the facility is open to the public by arrangement. Nearly 2,000 companies are represented in the archive's holdings, which include extensive collections from Marion Power Shovel, Euclid, Galion, Terex, Bucyrus-Erie, Caterpillar, and others. The museum houses over forty machines, some of which are fully restored and operable, and also includes a complete restoration shop.

If you want to become a member of this growing organization or need more information about the Historical Construction Equipment Association, visit our website at www.hcea.net, call (419) 352-5616, or email at hcea@wcnet.org.

Acknowledgments

The Historical Construction Equipment Association wishes to thank the following members for their help in bringing this book to fruition: Harry E. Young, for spearheading the selection of photographs; Roger Amato, Sam Robey and Dave Geis, for researching and writing the captions; Keith Haddock, for reviewing and editing the captions for historical accuracy and for use of information published in his books on mining machinery; Dr. John Thompson, for assisting in writing the captions for the dredges; and Thomas Berry, for scanning the photographs and editing.

MARION POWER SHOVEL COMPANY
CORPORATE HISTORY
By Keith Haddock

Henry M. Barnhart, Edward Huber, and George W. King incorporated the Marion Steam Shovel Company in Marion, Ohio in August 1884. A successful prototype steam shovel had been designed and tested by Barnhart a year earlier, and with the support of Edward Huber, sometimes known as the "father of Marion industry," manufacturing commenced in Huber's existing shops. After only three shovels had been completed, the young company was able to move into its own new factory in Marion, a site that would remain the company's home for the rest of its life.

New products soon emerged from the fledgling company. A land dredger known as *King's Ditcher* appeared in 1885. Invented by George King, the machine consisted of a half-swinging boom and dipper mounted on a land boat. Working from within the ditch, the machine dragged itself along by means of an on-board winch and cables attached to anchors in the ground.

Railroad construction formed the greater part of all excavation work in the 19th and early 20th centuries, and the products manufactured by Marion during this period were geared to this market. However, dredges began to play an increasingly important role in the company's product line in the early part of the 20th century. The experience and reputation gained from the ditching dredges produced in the late 1800s allowed Marion to expand its dredge line into many different types and sizes. Elevating dredges were introduced in 1904 and several were produced up to 1916. The world size record for a dredge was claimed in 1911 when Marion sold a 5-yard dipper dredge, and another record was broken the following year when the company built a massive electric placer mining dredge for the Klondike Gold operations in Canada. Marion also built a number of 14-inch and 16-inch cutter suction dredges starting in 1914. The dredge business remained a prominent line for the Marion Steam Shovel Company until the end of the 1920s.

Then the Great Depression and a saturated market took their toll on the dredge order books. However, Marion did not stop building dredges altogether—they continued building them to special order.

Marion's steam shovels progressed from Barnhart's early design to a full range of shovels by 1888. The company's steam shovel production received a boost during the construction of the Panama Canal when no less than 24 Marion shovels were shipped to that prestigious project. Like the original Barnhart, all steam shovels up to this time were of the railroad type; i.e. they ran on standard-gauge railroad track and their booms and dipper attachments were capable of swinging only about 180 degrees.

Marion launched its first fully revolving shovel in 1908, but at first this type of shovel enjoyed limited popularity because of competition from the heavy-duty, robustly designed railroad shovels of the day. In the early 1920s, revolving shovels rapidly gained in popularity as a result of improved designs and the innovation of multiple attachments capable of being used on the same base machine. These machines, known as "universal excavators," were generally made in sizes up to 2 1/2 cubic yards. They could operate as a shovel, backhoe, skimmer, dragline, clamshell, magnet, pile driver or crane. All of these functions could be performed by simply changing an attachment on the base machine and making a slight change in machinery arrangement.

In 1911, Marion put the first long-boom stripping shovel to work in North America. This was the Model 250 with a working weight of 150 tons, and a dipper capacity of 3 1/2 cubic yards. Even at this early date, Marion had already pioneered the concept of a hydraulic leveling cylinder under each corner of the machine, a feature found on every Marion stripping shovel built since that time. The 250 was followed by a succession of giant stripping shovels, many breaking the world record for size. Record breakers included the 8-yard 350 in 1923, and the 15-yard 5600 in 1929. Marion's revolutionary "knee-action" crowd first appeared on a 35-yard Model 5561, another record-breaking stripping shovel in 1940. That same year, the knee-action crowd was also applied to the 8-yard Marion 4121 coal-loading shovel. In April 1946, the company changed its name to the Marion Power Shovel Company to more closely reflect its products.

The year 1956 saw the first 65-yard bucketfuls from the famous Marion "*Mountaineer*," the first of the "super strippers" and arguably the most famous. Marion soon became famous for its stripping shovels, and these behemoths helped the company grow into one of the foremost manufacturers of excavating machinery. Marion's vast excavator range extended from the smallest 1/2-yard shovel to the largest ever put to work, the famous Captain Model 6360 stripping shovel in 1965 at the Captain Mine, Illinois. The *Captain* is reputed to be the heaviest mobile land machine ever built. Its huge dipper could carry 180 cubic yards of rock and earth!

Marion shipped its first walking dragline in 1939, and just three years later the company produced the world's largest dragline up to that time. More record-breaking draglines followed in 1963 (85-yard 8800), 1966 (130-yard 8900), and 1967 (145-yard 8900).

Marion made headlines when it built the famous Apollo moon rocket transporters for NASA in 1965. Based on stripping shovel undercarriage technology, the two diesel-electric transporters have taken part in most of the major space programs, including the Space Shuttle, and are still in use today. The 3,000-ton diesel-powered vehicles are designed to move fully assembled lunar spacecraft and rockets from the assembly building at Cape Canaveral to the launch pad three miles distant.

Marion introduced its first blast hole drill in 1967 following a joint development project with Peabody Coal Company. The Mark I drill could put down two 9-inch holes at the same time. Marion's drill line soon expanded to a variety of sizes up to 120,000 pounds pulldown.

Marion briefly entered the mining hydraulic excavator business with the introduction of the 3560 in 1981. However, this 300-ton, 20-yard machine gained only marginal success with only eight units sold.

In 1997, Marion Power Shovel Company was purchased by archrival Bucyrus International, Inc. (Formerly Bucyrus-Erie Company). The coming together of these two giants was a significant event in the earthmoving industry, and abruptly ended an intense competitive rivalry lasting 113 years. The plant at Marion, Ohio closed, but certain machines from the former Marion line have been updated and are still available as Bucyrus machines.

Editor's Foreword: The photographs in this book are part of the Marion Power Shovel Collection of the Archives of the Historical Construction Equipment Association in Bowling Green, Ohio.

This book is arranged into sections, based on the type of machine presented. The first section discusses shovels, starting with the conventional shovels and draglines that saw use in both construction and mining, followed by larger shovels and draglines built mainly for mining use, then the knee-action stripping shovels and finally the Superfronts. Large mining draglines, built specifically as draglines (as opposed to the convertible machines), from the early rail- and crawler-mounted draglines to modern walking draglines, follow the shovels. Dredges are next, and we'll finish with some of Marion's lesser-known products.

Some of the shovels shown here are also shown in our first book, *Marion Construction Machinery 1884-1975 Photo Archive.* This volume covers mining and quarry operations, and these machines are shown in those applications in this book.

Due to space limitations, this book is not intended to be comprehensive in its coverage. Instead, it shows representative machines of each type, with an emphasis on the older models that other books often do not cover.

This chain-hoist, half-swing, steam-powered railroad shovel predates the use of cable. The shipper shaft, to which the bucket is mounted, is powered by chain from the boom tip shaft at the tip of the boom to force, or crowd, the bucket into the bank. This design was referred to as half- or partial-swing because the boom swung roughly 180 degrees on a turntable mounted on the machine's frame, as opposed to the full-revolving design in which the entire upper works revolved 360 degrees on its base. The manually operated mechanical outriggers used on such machines are readily visible.

This wonderful machine, a Model C Tractor Shovel, worked for the Simons Brick Company of Los Angeles, California in 1906. It shows the use of traction wheels instead of rails and flanged wheels for travel; although it is free of rails, imagine its turning radius! As on the previous machine, the dipper stick crowd is driven from the boom tip shaft.

Marion produced 96 of these 5-yard Model 92 shovels between 1909 and 1930. This steam-powered, railroad-mounted version is seen working for The Wagner Quarries Company near Sandusky, Ohio in 1924. Marion started offering crawler mountings for its railroad shovels in 1923.

With a bucket capacity of six cubic yards, the Model 100 was Marion's largest railroad shovel. Between 1909 and 1926, Marion built 39 of these behemoths. This one is owned by Illinois Brick Company, of Chicago. The "Special" designation usually indicated a customization such as a longer boom or greater horsepower.

A Model 41 railroad shovel is seen digging clay for the Nova Scotia Clay Works at Pugwash, Nova Scotia in 1913. This 1 1/2-yard shovel has an independent steam crowd engine. Seventy-four were built between 1912 and 1923, and one more straggler was built in 1929, well towards the end of all railroad shovel production.

This impressive 4-crawler, 3 1/2-yard Model 70 electric shovel is a good example of the adaptation of the railroad-mounted, half-swing shovel to crawlers. Built between 1912 and 1930, 74 were produced.

Easy there, Dobbin! Besides exemplifying the occasionally spectacular disparities between loading and hauling technologies in the early 1900s, this Model 35 steam shovel illustrates the grousers that could be attached to steel traction wheels. Not a strong seller, only 53 of this model were built between 1908 and 1912. It takes a little finesse to dump 1 1/4 cubic yards of rock into that anachronistic and rather frail-looking cart!

This Model 28 is hard at work for the Woodland Clay Company in Woodland, Illinois in 1911. Marion built 402 of these 5/8-cubic yard shovels between 1911 and 1919. The smaller shovels, including the Models 21, 28 and 31, could be mounted on steel traction wheels, flanged wheels for use on rail, or crawlers, which were first offered in 1915.

A one-yard Model 31 steam shovel is seen loading coal for T. J. Forschner Coal Company at Linton, Indiana. Marion produced 279 of these machines between 1912 and 1923. The Model 31 was the first Marion shovel to be offered with diesel power, and a channel-type boom was available for crane applications. While widely used in construction, conventional shovels like this also saw extensive use in mines and quarries.

This rather complicated arrangement must be one of the earliest "long crawler" designs. The Model 36 could also be ordered with wheels or on rails. Replacing the Model 35, it was a 1 1/2-yard machine with 253 built between 1912 and 1923.

A steam-powered, 3/4-yard Model 21 is loading clay in 1925. This was a very successful machine for Marion, with approximately 810 units produced between 1919 and 1926. Like its counterpart Models 28 and 31, it was offered with steam, gas, or electric power. An electric-powered Model 21 is preserved at the Historical Construction Equipment Association's National Construction Equipment Museum in Bowling Green, Ohio.

That fireman standing at the door may be contemplating the loss of his trade to electric and internal combustion power on shovels. This 1 1/2-yard Model 32 steam shovel is one of 229 Model 32s delivered between 1922 and 1933, with a final unit shipped in 1941.

A Type 490 electric shovel with a 2 1/4-yard bucket is digging limestone in 1927. The power cable and the large, exposed drive gears for the crawlers are apparent. Marion built 65 of these crawler machines between 1926 and 1937, several of which were sold to the Six Companies joint venture for the construction of Boulder (now Hoover) Dam.

This very subtly lettered Type 460 electric shovel is shown in 1927. The delicately balanced boulder demonstrates the smooth operation of its Ward-Leonard electric controls. Twenty-six of these machines were built between 1927 and 1934.

The Type 480, introduced in 1928, was a 2-yard shovel. By 1944, its last year of production, Marion had delivered 140 units. This diesel-electric version was photographed in 1928.

In August 1935 a Type 480 shovel loads coal for the Delta Coal Company. As with earlier models, the 480 was offered with steam, diesel or electric power. The five men are cleaning up the loading area with hand shovels.

Marion's introduction of the Type 93-M in 1946 began a very popular run of 364 units produced through 1975. It was rated at 2 1/2-cubic-yard shovel capacity, a very popular size for heavy excavation from the 1940s through the early 1960s.

Mullet Coal Company of Mt. Eaton, Ohio puts their Type 111-M dragline to work removing overburden as a dozer assists. The Type 111-M was offered from 1946 to 1974, with 379 units sold. Like its counterparts, the Type 93-M and the later 101-M, the 4 1/2-cubic yard 111-M was offered with diesel or electric power.

Loading shot rock in the pit with this Type 111-M electric shovel went well as long as the power cables were in good condition and were kept out of the water as shown. Marion introduced electric-powered 111-Ms with Ward-Leonard controls in 1948. In the 1960s, shovels in the 4- to 5-yard range saw wider use in general construction and quarrying applications, replacing many smaller machines.

The 3-yard Type 101-M was introduced in 1953. One hundred twelve were sold, largely for export. By the early 1970s, hydraulic excavators and large wheel loaders were rendering most cable excavators of 5-yard capacity or less obsolete. This obsolescence was a factor in Marion's decision in 1975 to discontinue its construction product line and concentrate on large mining shovels and draglines. The Types 93-M and 101-M were the only models in the line still in production, the sole survivors of a long and rich lineage of construction shovels.

This 2-yard Type 80-M, built in 1960, is unique. Only one was built, and here it loads an Euclid dump truck for the Maumee Stone Company in 1961. The 80-M was derated to 1 3/4-yards capacity and redesignated as Type 75-M.

A new 4-yard Type 4160 electric shovel occupies several rail cars on its way to The Marble Cliff Quarries Company. Compare this fairly compact load to the later machines, which could require 100 or more railroad cars for transport!

The same Type 4160 has been assembled and is now working for Marble Cliff Quarries in 1927. Marion built 65 units between 1927 and 1937. Note the relatively short boom and dipper stick; as with modern excavators, a short boom and stick allows the use of a larger bucket for greater capacity but gives up some of the reach, digging depth and dumping height.

This very unusual Type 6200 electric shovel is seen working in the Mesabi Iron Range of northern Minnesota for Pickands Mather & Company. Only two of these enormous 5 1/2-yard shovels were produced, both in 1929. The elevated operator's cab and the rather obsolescent half-swing design were unusual for the era.

Offered from 1930 through 1933, only 13 of the 3-yard Type 4120 electric shovels were sold, probably due to its untimely introduction during the Great Depression. This example is loading a quarry train pulled by a Plymouth locomotive. Narrow-gauge trains saw widespread use in mining and quarry hauling applications like this until large off-highway trucks became practical.

A 6-yard Type 4161 electric shovel loads Mack LRSW 30-ton quarry trucks for the Interstate Iron Company. That sled in the bottom of the picture is a rather untidy method of dealing with the power cable.

The Type 4161 was a very successful shovel, with 206 units delivered between 1935 and 1964. Marion changed its shovel front design from single dipper sticks to twin sticks to better transmit load-related stresses from the dipper to the boom. This twin-stick 4161 is loading an Euclid 1LLD twin-engine, 50-ton dump truck.

Introduced in 1945, the 151-M electric shovel had a rated capacity of 7 1/2-cubic yards and was manufactured as recently as 1982, with 170 units produced. Electric shovels of this size were used most often in mines and quarries, but some also saw use in heavy construction.

This later version of a 151-M shovel is put to use loading haul trucks, probably in a pit with less overburden. Note the large screened area of the house to allow airflow for the radiators of the two 350-horsepower Cummins diesel engines.

In 1951 Marion introduced the 10-yard 191-M, claiming the title for the world's largest shovel on two crawlers. Later improvements more than doubled its rating to 22 cubic yards, and its demand required production to continue until 1989. Most were electric, but a few were built with diesel power. Several were built as draglines.

After having used Marions at Hoover Dam as part of the Six Companies joint venture, construction giant Morrison-Knudsen of Boise, Idaho is known to have owned only eight Marion machines in the post-war years. This 1969 Type 191-M electric shovel is shown loading a Haulpak end dump at a coal mine at Kemmerer, Wyoming, where M-K had contracted the overburden removal; it was shipped here after starting its career at M-K's Cedar Springs Dam project in southern California. The other Marions in M-K's fleet were an 8200 dragline and six construction-line machines: a 93-M for Venezuela in 1951, a 111-M delivered in 1948 to a project in central Washington and four more 111-Ms sent to Vanderhoof, British Columbia in the early 1950s.

Bethlehem Limestone Company of Hanover, Pennsylvania uses their Type 181-M electric shovel to load a fleet of well-maintained Euclid tractors with belly dump coal wagons. The 181-M was first offered in 1956, with 21 units from 8- to 16-yards capacity produced. Manufacturing of the 181-M ended in 1977.

A throwback to Marion's rich history as a dredge manufacturer, this barge-mounted Type 183-M clamshell exemplifies how Marion worked with its customers to provide the best solutions for any type of material removal. Built for Mall, Inc. of Baton Rouge, Louisiana, it was nicknamed the *Big Mall*. Although the three large spuds were required to hold the barge in place as the clam worked, they reduced the 183-M's ability to rotate 360 degrees. Marion built 37 183-Ms, all diesel-powered, between 1956 and 1974. The standard 7- to 10-yard capacity 183-M was crawler-mounted and available as a dragline or long-boom stripping shovel.

In 1962, Marion again broke the world size record for a 2-crawler shovel when it delivered the first of two 15-yard 291-M long-range shovels to Peabody Coal's Sinclair Mine in Kentucky. A year later, the second was delivered to the same company's Lynnville Mine in Indiana. Designed to remove parting material from between two coal seams, their long 90-foot booms were able to cast the material well clear of the pit. The 291-Ms enjoyed an extremely successful career. The Sinclair machine later worked a few years in Oklahoma and, in the mid-1980s, both 291-Ms were moved to Peabody's North Antelope/Rochelle Complex in Wyoming where they worked as 40-yard coal-loading shovels until 1998.

English contractor George Wimpey and Company, one of the world's largest construction companies, is loading Euclid R-45 haul trucks with a 182-M electric shovel at their Bryn Pica Coal Mine near Aberdare, South Wales in 1979. One has to notice the wonderful office-like control cab that Marion has provided. Capacity of the 182-M ranged from 9- to 18-cubic yards. Introduced in 1966, it was still offered when Marion ceased operations in 1997. Fifty 182-Ms were sold.

Here's another big Marion at Wimpey's Bryn Pica operation, a 195-M dragline. Equipped with a 130-foot boom and 12-cubic yard bucket, it removes overburden above the coal seams and loads it into haul trucks at the rate of 5,000 cubic yards per 16-hour day. The 195-M was offered only as a dragline, and 18 were sold between 1970 and 1993. Bucket capacity went as high as 17 cubic yards.

The Waterloo Coal Company puts their Type 184-M to use removing overburden at one of their open pit coal mines. The 184-M was offered only as a dragline, powered by a diesel engine. Fourteen of the 8- to 12-cubic yard draglines were manufactured between 1973 and 1982.

This Type 201-M electric shovel is working for the Martiki Mine in Kentucky. The 201-M was rated at 16 to 35 cubic yards; 50 were produced between 1975 and 1989.

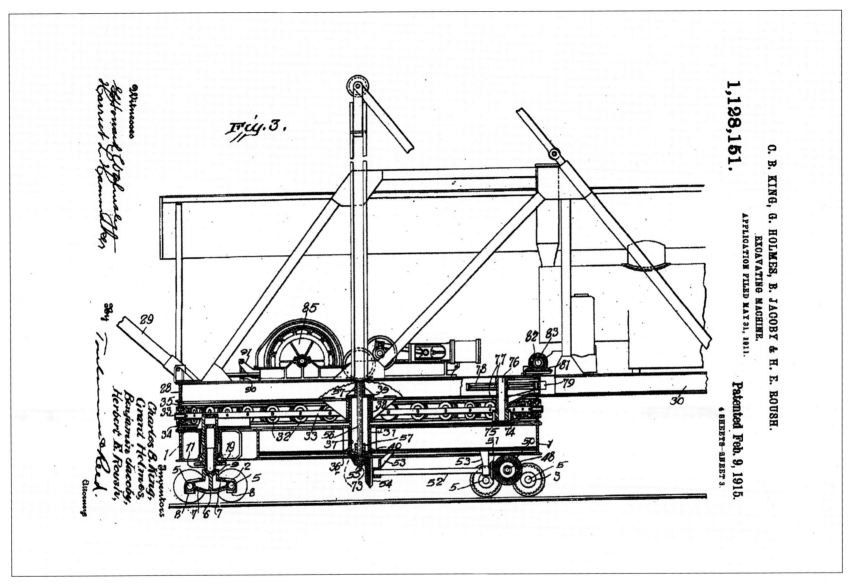

C. B. KING, G. HOLMES, B. JACOBY & H. E. ROUSH.
EXCAVATING MACHINE.
APPLICATION FILED MAY 31, 1911.

1,128,151.

Patented Feb. 9, 1915.
4 SHEETS—SHEET 3.

The Marion 250, introduced in 1911, was the first full-revolving long-boom stripping shovel designed and built in North America. It swung a dipper of 3 1/2 cubic yards on a 65-foot boom. This patent drawing illustrates the hydraulic leveling cylinders that Marion pioneered on the Model 250 and used on all subsequent models. Using a cylinder in each corner of the machine, this system leveled the machine and compensated for uneven working surfaces in mines and quarries.

The first Model 250 was put in service by the Mission Mining Company near Danville, Illinois in 1911 and was retired in 1934. The dignitaries taking a close view of the 3 1/2-yard bucket include W. G. Hartshorn (extreme left), the customer who persuaded a reluctant Marion Steam Shovel Company to build this pioneering machine, and Grant (Jack) Holmes (second from right), designer of the hydraulic equalizing cylinder for such shovels.

The Model 250 had a 65-foot boom and an operating weight of 150 tons. The efficiency introduced by the Model 250 made surface mining practical. It was well received by customers, with 13 delivered between 1911 and 1915.

Only one Model 210 was built, a 2-yard full-revolving railroad-mounted shovel sold to the Fulton Brick Company in 1912. The tracks on which mining shovels traveled were often quite rough, and the slightly uneven rails in this picture amply demonstrate the advantage of Marion's hydraulic leveling cylinders.

Building on the success of the Model 250, Marion quickly designed a substantially larger shovel, the Model 270, that was introduced in 1912. Five of these 5-yard machines were built before it evolved into the Model 271 in 1913. This print was retouched to show dimensions for reach and digging depth.

A 2-yard Model 211 at work for the Ohio Valley Coal Company. The Model 211, a very large machine for its time, had an operating weight of 95 tons. Offered between 1913 and 1915, only two of these machines were built. Rather than reflecting on the quality or success of the machine, the typically low numbers of large shovels and draglines sold is usually indicative of the very limited demand and market for which they were built. Many were built on a special-order basis.

The Model 251 was introduced in 1913 as a 3 1/2-yard successor to the Model 250. The Model 251 weighed 185 tons, and 10 had been built by the time production ended in 1916. This example, working for Liberal Coal Company at Liberal, Missouri, illustrates how these early stripping shovels revolutionized surface mining.

A Model 271 strips overburden near Joplin, Missouri for the Minden Coal Company. Between 1913 and 1917, Marion delivered nine of these 5-yard shovels. The Model 271 had an operating weight of 213 tons. With the shovel facing the appropriate direction, that coal sled attached to the shovel by cable would be within shoveling distance of the house's back door.

The track crew is busy laying ties and rails to enable this Model 271 to advance further into the cut. Crawler track assemblies at each corner would eventually eliminate this inefficient step in strip mining. The twin stacks indicate the presence of a second boiler to meet the big machine's power requirements. The Model 271 was also available with electric power which, while cleaner to run, certainly detracted from certain aesthetics!

A 6-yard Model 300 is seen loading an ore train. Introduced in 1915, this 350-ton monster was available with either steam or electric power. It was very successful, with 74 produced by the time it was superseded by the Model 350 in 1923. In 1919, the Model 300 became the first Marion electric machine to use the newly developed Ward-Leonard controls, which greatly improved the performance of electric shovels.

The Model 252, introduced in 1916, evolved from the Models 250 and 251. While its predecessors were available only with steam power, the Model 252 offered electric power as an alternative. Ten of these 200-ton, 3 1/2-yard shovels had been delivered when the last was shipped in 1927. Note the change in boom design; the Model 252 used an enclosed, fabricated boom, as opposed to the lattice booms of the Models 250 and 251. This steam-powered Model 252 is loading clay for the Back Brick Company of Chicago in 1925.

The Model 350 was a world record machine when introduced in 1923 with its 8-yard bucket, 90-foot boom, and an operating weight of 560 tons. It was the largest steam-powered shovel built by Marion; the men on the ground in this heavily retouched photo give a sense of its size.

This electric Type 350 was very likely built after 1926, when Marion replaced the word "Model" with "Type" in its lettering. Marion started mounting the Type 350 on crawlers in 1925, freeing the stripping shovel from rails. Between 1923 and 1929, Marion sold 35 of these large machines, but records indicate that one Type 350 was shipped to Russia in 1943! One, built in 1927, is preserved at the Diplomat Mine Museum, at Forestburg, Alberta, Canada.

The Model 125, introduced in 1925, was a 3-yard machine mounted on four crawlers. It was available with either steam or electric power. By the time production ended in 1933, 22 shovels and 16 draglines had been delivered. Note the jib for hoisting coal to the firebox.

A Marion Type 5120 being erected at the factory in May 1929, the year of its introduction. The 5120 was a 3-yard electric shovel that could also be ordered with diesel-electric power. Five of these machines were built between 1919 and 1935.

In 1927, both Marion and its arch-competitor Bucyrus (which merged with Erie Shovel to create Bucyrus-Erie that same year) introduced machines rated at 12 cubic yards capacity. Marion's machine, designated the Type 5480, had an operating weight of 975 tons. The all-riveted construction of the steel structural members would soon give way to electric arc welding. That network of frame members serves to distribute the stress of the boom to the frame of the upper works. Later models would have this massive balanced design fully enclosed in the house's sheet metal. Notice also the cable reel, which was developed to more efficiently handle the electrical power cable. Eleven Type 5480 shovels and four draglines were produced between 1927 and 1932.

The Type 5320 electric stripping shovel was introduced in 1929. Six units were built through 1934, when the model was discontinued. Bucket sizes ran from 8 to 12 yards, and operating weight was 725 tons.

The Type 5600 is another Marion shovel that broke records for size and production when introduced in 1929. Only one of these 15-yard, 1,550-ton shovels was built. This shovel exemplifies the upper limits of shovel design prior to the introduction of knee-action crowd in 1940. Note the smaller Marion shovel, a Model 480, loading coal into the rail cars as a Heisler steam locomotive waits to set the next empty car for loading.

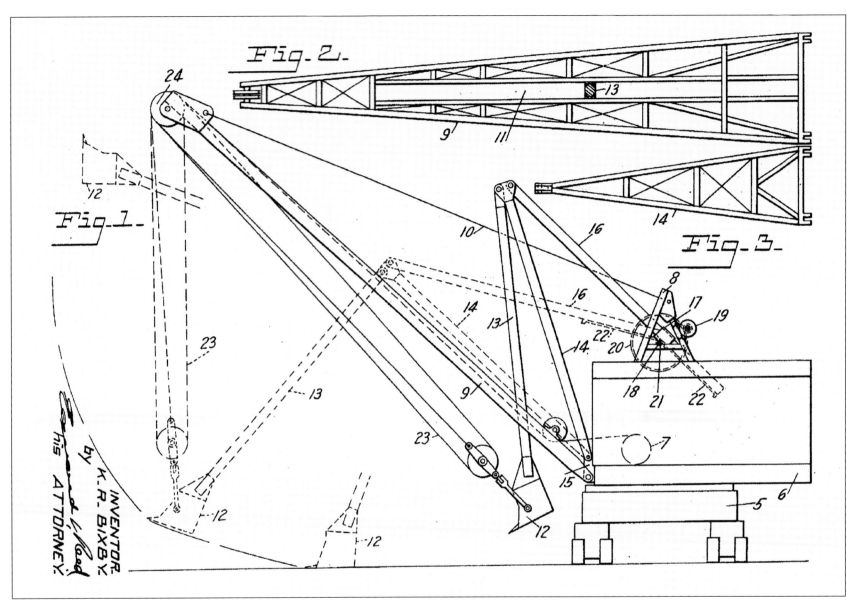

Like the later Superfront design for smaller machines, knee-action crowd on the larger stripping shovels provides the following: increasing digging power as the bucket is crowded into the bank, horizontal bucket motion into the cut (which protects the coal beneath the overburden being removed), more efficient swing, and elimination of bending stresses. The basic knee-like movement of the dipper is shown in this patent drawing from 1938.

Here's an early 1950s view of a 35-yard Type 5561 shovel. Introduced in 1940, it was the first shovel with Marion's revolutionary knee-action crowd mechanism, and the biggest stripping shovel built to date. The company sold 17 of these 8-tracked, 1,675-ton machines through 1956.

This 8-yard Type 4121 electric shovel, one of 27 built between 1934 and 1949, was the only conventional shovel (as opposed to large stripping shovels) equipped with knee-action crowd. Built in 1940 for loading coal, it works at the Hawthorne Mine in Indiana. Though there was no direct official correlation, it was in some ways the predecessor to the Superfront shovels of the 1970s and 1980s in that the Superfronts, also derived from conventional shovels, incorporated the same horizontal bucket motion to improve digging efficiency. The truck, by the way, is a Mack FCSW tandem chain-drive, towing a gravity side-dump trailer.

A dipperful of overburden takes the big plunge as a Type 5323 with a 160-foot plate-and-lattice boom works for Marquette Cement Company in Ohio. This was a special long-boom version carrying an 11-yard dipper. Bucket capacities of the 5323 normally ranged from 15 to 20 yards, and operating weight was just over 1,000 tons.

This Type 5323 has a steel plate boom. Although Marion offered the 5323 from 1941 to 1961, only nine machines were built, illustrating the limited market for these huge shovels.

This is the 8-track base of the Type 5323 stripping shovel during erection. Note the stiff-leg derrick on the bank; large cranes were standard equipment for assembling electric shovels, stripping shovels and walking draglines on site.

As the first of what became known as the "super strippers," this Model 5760 heralded a new era in mining excavator evolution. Named *The Mountaineer*, the 65-yard shovel was built in 1955 and worked in the eastern Ohio coal fields for the Hanna Mining Division of Consolidation Coal Company. *The Mountaineer* had a 160-foot long boom and a total weight of 2,750 tons.

A 1950s view of a Model 5760 at the Midland Electric Coal Company's Farmington Mine near Victoria, Illinois. Marion erected five Model 5760 shovels between 1955 and 1959; with bucket capacities to 70 yards offered for the 5760, this 60-yarder was the runt of the litter.

This Model 5761 is seen opening a new highwall with its 65-yard bucket in a surface mine. The 5761 was introduced in 1959 as a replacement for the Model 5760, and 16 were delivered before discontinuance in 1970. The stiffleg upon which the dipper stick pivots on the knee-action crowd shovels is easily seen. While the dipper is raised by cable, the crowd and backhaul action is powered by machinery in the A-frame structure atop the shovel's house. The stiffleg pivots on the frame of the shovel itself, moving back and forth with the dipper stick as it is crowded forward and backhauled.

Here is Marion's 5860 in action. The 80-yard stripper has a 180-foot boom and weighs 5,075 tons. Marion built two of these stripping shovels in 1965 and 1966. Notice that the knee-action dipper stick passes freely through the boom to its pivot point atop the stiffleg. On a conventional shovel, it passes through a pivot in the boom. A crowding arm connects the stiffleg and dipper stick to the crowd and backhaul machinery in the A-frame.

Weighing in at approximately 15,000 tons, the Model 6360 was, along with Bucyrus-Erie's famous 14,500-ton walking dragline *Big Muskie*, one of the two heaviest land-based mobile machines ever built. The only 6360 built, it was named *The Captain* after the Southwestern Illinois Coal Corporation mine near Percy, Illinois at which it worked.

The 6360 had a massive 180-cubic-yard bucket and a purchase price in 1965 of approximately $15,000,000. The pickup truck and wheel loader next to the crawlers and the man in the circle next to the bucket give a sense of *The Captain's* enormity. Sadly, the great shovel suffered a severe electrical fire in 1991 and was scrapped the next year after having moved just over 800,000,000 cubic yards of overburden.

This 105-yard Model 5900 is working at Peabody Coal Company's Lynnville Mine in southwestern Indiana. The shovel was delivered in 1968. It has a 200-foot boom and weighs 7,250 tons. As of 2002, this machine was still intact but had been parked. One other 5900, with a 110-yard bucket, was completed in 1971. That machine, built for AMAX Coal Company, worked at its Leahy Mine in southern Illinois, and was later moved to the adjacent Captain Mine. It was the last stripping shovel built by Marion.

The only Model 5960 built is shown at the Peabody Coal Company's River Queen Mine at Greenville, Kentucky. The shovel, named *Big Digger*, weighed 9,000 tons and had a 215-foot boom and 125-yard bucket. It was completed in 1969 and scrapped in 1990.

While space doesn't allow a full explanation of Marion's revolutionary Superfront shovel, suffice it to say that it was based on two ancient laws of physics: The triangle as the strongest possible structural design, and the multiplication of force in proportion to the length of a lever to which the force is applied. This rather stylized artist's rendition illustrates its radical departure from conventional boom-and-dipper stick design.

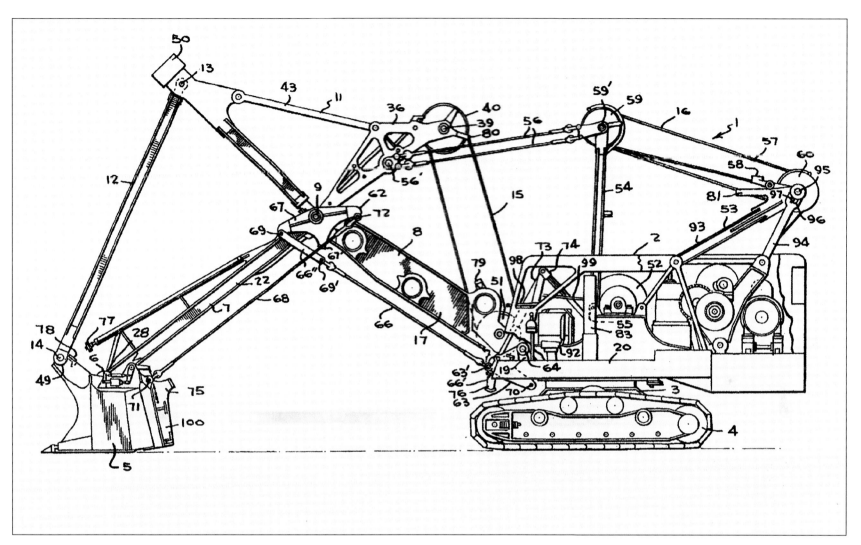

While the workings of a Superfront may look complex in this 1970 patent drawing, the principles behind it are simple: to dramatically increase bucket capacity by decreasing overall front end weight, largely by eliminating the boom and stick and adding the weight of a substantially larger bucket in their place; to increase digging force by allowing the bucket to be rotated by applied leverage as it moves through the bank to load; to eliminate bending stresses that the booms and sticks of conventional shovels undergo when loading by applying the principle of the strength of the triangular structural unit; and to increase machine swing efficiency by moving the overall center of gravity closer to its center of rotation.

The Superfront design originated in the mid-1960s with a machine designed to more efficiently excavate coal from thin seams by allowing a level digging plane as opposed to the arc typically described by the bucket of a conventional shovel. As its superior digging power became apparent, the Thin Seam Miner evolved into the Superfront for other digging applications. This 101-M, with house removed, was converted to a Superfront test bed in 1967; the design allowed for a 5-yard bucket instead of its standard 3-yarder by virtue of eliminating the weight of the conventional boom and dipper stick.

Two standard 15-yard 191-Ms were converted to Superfronts, with new 194-M model designations, in 1972. This one was a 16-yard machine for Reserve Mining Company's taconite mine in Minnesota.

The other 194-M was built with a 22-yard bucket for Cyprus-Pima Mining Company's copper mine near Tucson, Arizona. Here it's loading a Wabco Haulpak end dump.

The Superfront shovel was formally introduced in 1974 as the Model 204-M. All 204-Ms were electric, with bucket capacities from 20 to 40 cubic yards. The first order for 204-Ms was the largest, with ten machines shipped to the former Soviet Union from 1976 through 1981. This one is working for Kuzbass at the South Yatusk Project in Siberia.

Seven other 204-Ms were built, culminating with four sent to copper mines in Papua New Guinea in 1987 and 1988, the year they were discontinued. The other three went to Australia and to coal mines in Wyoming's Powder River Basin.

The Model 220 was a rail-mounted, steam-powered dragline with a bucket capacity of 2 yards. Five were built between 1913 and 1915. This machine is working for the Pierce Coal Company. With its four four-wheel trucks, it may be giving a hint at the arrangement of crawlers at the corners of large excavators several decades later.

Marion produced only three Model 261 steam-powered draglines. The 3 1/2-yard machines were all delivered in 1914. Winston Brothers Company owns this rail-mounted example. A railroad dragline's travel rails had to be pulled up as the dragline moved back from the cut, whereas a railroad shovel's track, once laid, could be left in place indefinitely. Crawler draglines, of course, required only reasonably level ground for travel.

The Type 360 was the dragline version of the Type 350 shovel. This electric-powered, 4-crawler machine is stripping overburden for Northern Illinois Coal Corporation in 1929. Between 1922 and 1928, twelve of these draglines were delivered.

The Model 4120 was a 3-yard, conventional electric dragline mounted on two crawlers. Thirteen were delivered between 1930 and 1933.

The 1 1/2-yard Type 361 dragline was offered between 1933 and 1938. A total of 41 Type 361s were built. They could be supplied with gas, diesel, or electric power, and could be mounted on crawlers or steel rails. Some early draglines traveled on rollers.

Marion's first walking dragline was the 7200, introduced in 1939. Walking draglines get their name from their traveling system. The shoes on each side are rotated up, backwards and back down, then the machine is repositioned back onto the shoes by the same mechanism, producing a sort of walking motion. The walking mechanism is part of the main frame of the house, and when not traveling the shoes are raised and the machine rests and revolves on a tub-like base. This 7200, owned by the Dunn Fuel and Lumber Company of Coalgate, Oklahoma, is equipped with 135 feet of boom. The 7200 draglines handled 5- to 8-yard buckets, and 57 were built with diesel or electric power between 1939 and 1958.

In 1957, the Pittsburgh Coal Company is removing overburden with their 7400 at one of their mines. Bucket capacity on the 7400 ranged from 7 to 14 cubic yards, and boom lengths from 160 to 235 feet. With 92 machines built over 35 years between 1940 and 1974, the 7400 set records for Marion walking draglines both for the numbers built and the time span over which it was produced.

The Aluminum Company of America (ALCOA) uses this 7800 dragline to mine bauxite in Rockdale, Texas. When the 7800 was launched in 1942, it claimed the title of the world's largest dragline. It had bucket capacities of 20 to 35 cubic yards. Nineteen 7800s were assembled between 1942 and 1964. As the size of these draglines increased, bridge cranes with capacities of 100 tons or more began to appear on the draglines. The crane travels on tracks along the ceiling of the house; these tracks protrude from the back wall of the house, as seen on this machine. By opening doors on the back of the house, the crane can be used to install, service and remove major components on the machine deck, passing them out the back of the dragline.

This 7900 walking dragline owned by Utah Construction & Mining Company is being relocated from one area of a mine near Farmington, New Mexico to another. The land over which the dragline is moving has been leveled and graded to insure safe travel. A walking dragline must always travel backwards, opposite of the control cab. To steer or turn, the operator simply raises the walking feet, swings the machine inline with the new direction he wishes to travel and then begins the walking process again. A relatively rare machine, only six 7900s were produced over ten years between 1962 and 1971.

Marion broke the world dragline size record by a large margin in 1963 when it introduced the massive 8800 with an 85-yard bucket on a 275-foot boom. Weighing in at over 12,500,000 pounds, it was rated at 10,950 horsepower. The only 8800 built is seen at work for the Peabody Coal Company near Beaver Dam, Kentucky where it later operated with a 100-yard bucket.

Only two Model 8900 walking draglines were built, and Peabody Coal Company purchased them both. The first one went to its Moura Mine in Queensland, Australia in 1966, the second went to its Dugger Mine in southern Indiana in 1967. The first 8900 had a 130-yard bucket, setting another world record. The second machine broke that record with a 145-yard bucket on a 250-foot boom, and was upgraded with a 155-yard bucket in 1993. The 8900 had an operating weight of 7,000 tons.

Marion built three 8400 draglines between 1969 and 1971. This model, covering the 60- to 80-cubic yard range, was overshadowed by Marion's greater-capacity 100-yard class 8750, which was introduced in 1971. This was followed by Marion's extremely successful 8050 and 8200 models introduced in 1972 and 1973 respectively. These two models, which spanned bucket sizes from 50 to 94 cubic yards, amassed total sales of 69 units.

Marion took a different direction in the boom design for this Model 7500 dragline. Although Marion nearly always used I-beam and H-beam structural members as opposed to tubular steel in the construction of their booms, on this machine they used many more vertical and horizontal members and greatly reduced the amount of lattice bracing. The 7500 could handle up to a 20-cubic yard bucket; 16 were built between 1970 and 1981, all with electric power.

This 7820 dragline is hard at work for Vantage Coal Corporation near Clarion, Pennsylvania. The 7820 was supplied buckets from 32 to 44 cubic yards, and was electrically powered. It was in production from 1970 until 1993, and 13 were built.

Optimum Collieries Company of South Africa named their Model 8000 walking dragline *Thor*. Only two of this model were built, in 1970 and 1971. Both came with 55-cubic yard buckets and 275 feet of boom. A spare spreader beam and wire rope sheave are visible in the foreground. Owners of large draglines kept at least one extra bucket on hand. With one in use on the machine, the other would be reconditioned in the shop. Some owners kept a third bucket in reserve and cycled the buckets through duty on the machine.

Utah Construction and Mining's 8050 walking dragline was produced in the "boom years" for large draglines, with 35 machines being built between 1972 and 1986. Bucket capacity ran from 51 to 64 cubic yards. If you can see the man silhouetted under the back of the dragline, you can appreciate the unbelievable size of these machines. Utah operated eight 8050s, with one in New Mexico and the rest in Australia.

The largest dragline ever produced by Marion was its 8950 model, equipped with a 310-foot boom and a 150-cubic yard bucket. The only 8950 ever constructed was purchased in 1972 by the Amax Coal Company and shown here working at their Ayrshire Mine near Evansville, Indiana. With each cast, the big dragline could relocate over 225 tons of rock and earth to a point over 600 feet away.

Marion's draglines were often referred to as "life of mine" machines. They were designed for a single location and use. The machine was designed, engineered, built, assembled and operated at a single location for its entire life, up to 35 years. Here's Morrison-Knudsen Company's 8200, the *Crows Nest*, at work swinging its 75-yard bucket at Hardin, Montana. Thirty-four Model 8200s, from 65 to 94 yards, were delivered between 1973 and 1996.

The only 7620 walking dragline built removes overburden for Knife River Coal Company near Gascoyne, North Dakota. Built in 1974, it has a 30-yard bucket. Notice the vast array of catwalks and ladders required on the boom to service and maintain the working components, as the boom was kept at a set angle during operation and would most likely never be lowered to the ground.

In a very unusual application for a walking dragline, Luscar Sterco of Canada is employing this electric-powered 7450 to excavate and load coal from a seam over 200 feet deep. The 7450 was offered with a bucket range of 10 to 14 yards and diesel or electric power. As the size of the draglines and the length of their booms increased, the base width of the boom had to be increased to handle the swing stress imposed on the connection between the boom and the frame. Seven 7450s were built between 1979 and 1985.

Dixie Fuels was the only customer that put one of these 7250 walking draglines to work. Completed in 1982, it was electric powered with an on-board twin Cummins diesel generator, and wielded a 13-yard bucket. The electric fans used to move air through the machine are very prominent on the roof. This picture also clearly shows the connecting rod system on the walking mechanism.

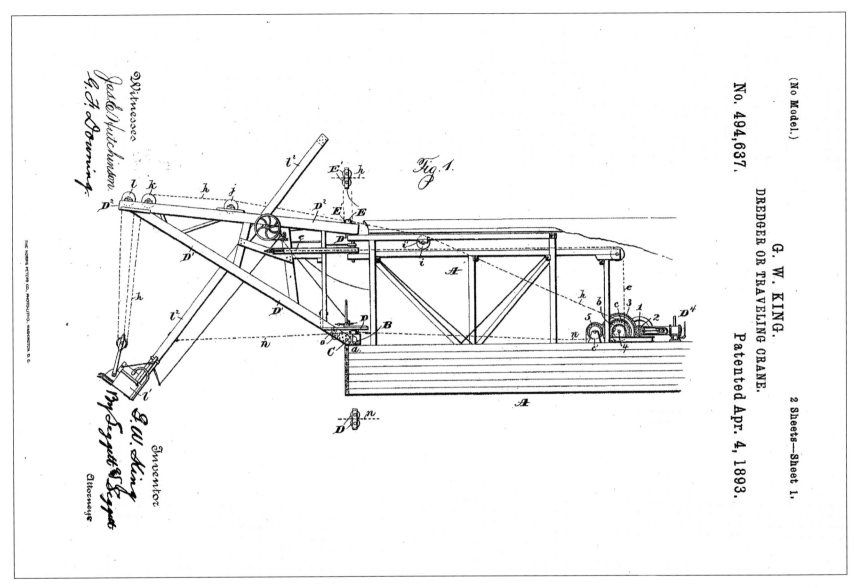

Fig. 1

George W. King, co-founder of Marion Steam Shovel, designed and built steam excavators ranging in capacity from 1/2-cubic yard ditchers to 5-cubic yard dredges, and electric-powered gold dredges with ladders carrying buckets of as much as 16-cubic foot capacity. Among the numerous patents issued to King was one for this crane-type dipper dredge in 1893. Nevertheless, most Marion dipper dredges carried the scoop in a boom that rode on a swing circle and was supported by an A-frame.

Marketed as the King's Ditcher, this 7/8-cubic dredge was probably the smallest steam excavator used in digging the Chicago Sanitary and Ship Canal. The 28-mile canal connects Lake Michigan with the Illinois and Mississippi rivers. Begun in 1892, the project took eight years to complete. The stout, wedge-shaped frame had a flat bottom about 6 feet across and a 16 by 30-foot platform for the boiler, 2-horsepower engine, and hoisting drum. To move the dredge, the spuds were raised and the dredge was winched forward from deadmen anchored to the ground ahead of it. King's Ditchers were marketed principally to drainage and irrigation system contractors and to developers of municipal aqueducts.

A 2 1/2-yard traction dredge is loading railroad dump cars on the Chicago Sanitary and Ship Canal project in 1892. Marion sold six or seven traction dredges between 1892 and 1904. The dredge carried a hoisting engine with two 10 by 12-inch cylinders, an 8 by 8-inch swinging engine, and locomotive boiler on a frame measuring 44 by 21 feet. The rated capacity of the rig was 900 to 1,800 yards per 10-hour day.

This 1 1/4-yard traction dredge was erected in 1904 in west central Illinois to build levees. The cable-operated machine had an 8 by 10-inch hoisting engine, a 6 by 7-inch swinging engine and a 55-foot boom. She was rated at 400 to 800 yards per 10-hour day, and the crew's quarters were on the upper deck.

Marion wasn't exactly subtle about publicizing who built this dipper dredge. Along with the boom support cables, the A-frame above the crew also supports a pair of detachable telescoping spuds. The spuds were extended against the banks of the canal to stabilize the dredge as she worked. Angers & Mestayer of Kaplan, Louisiana is the owner.

Marion made ditching machines of 1/2 and 3/4-cubic yard capacity, with each size being available as a dredge with steel pontoon or wooden hulls, a skid ditcher on runners, or a mobile ditcher on railroad-type trucks and portable tracks. These machines could dig between 200 and 800 yards respectively in a 10-hour day. This photograph of a ditcher mounted on railroad wheels is dated August 29, 1904.

The most common ditching dredges were of 1- to 2 1/2-yard capacities. Most had locomotive-type boilers with recommended working pressures of 125 pounds per square inch. Depending on boom length, the small rigs dug 350 to 800 yards per 10-hour day, while the 2 1/2-yard dredges dug between 700 and 2,000 yards. Boom lengths ranged from 32 feet for the 1-yard machine to 80 feet for the 2 1/2-yard dredge. The boom is swung by the cable-powered turntable on which it is mounted.

The hoist works for a 1 1/2-yard steam-powered dredge are seen inside one of the Marion factory buildings. The hoist works were completely assembled and tested at the plant before shipping, and all the components for each dredge were shipped as a lot for assembly on-site.

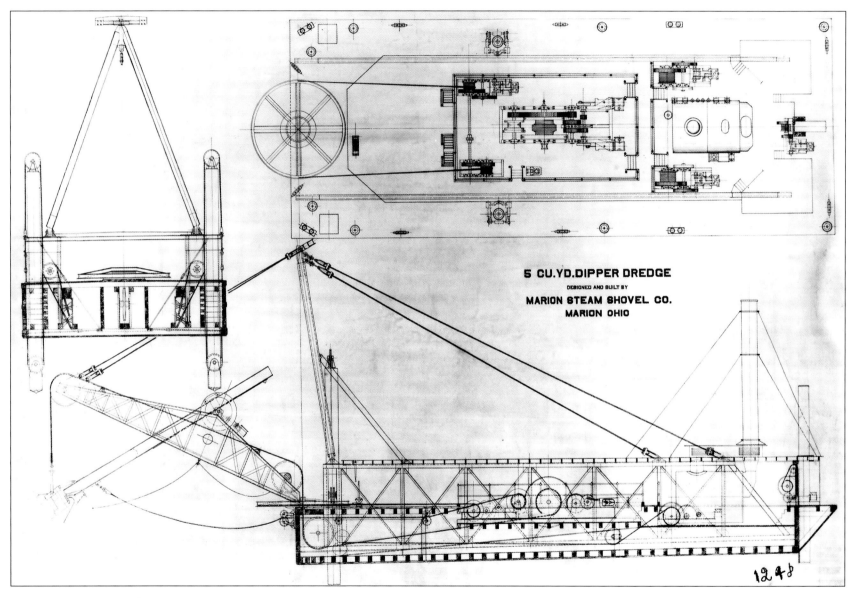

5 CU.YD. DIPPER DREDGE

DESIGNED AND BUILT BY

MARION STEAM SHOVEL CO.

MARION OHIO

A line drawing of the Marion 5-yard dipper dredge, which was introduced in 1911. The spuds that anchor the dredge in place while digging are also raised and lowered by steam.

The *Cyclone* was a 5-yard steam dipper dredge, the largest built by Marion. She was built in 1911 for James Stewart & Company, and is shown working on the New York State Barge Canal at Baldwinsville, 12 miles northwest of Syracuse.

Among the 18 dipper dredges of 4- and 4 1/2-yard capacity built by Marion was the *Kewaunee*. She was built in 1913 for the Milwaukee District of the U.S. Engineer Department (later the Corps of Engineers) at a cost of $47,750. A Scotch Marine boiler generated 115 pounds per square inch of steam for both the hoisting and swinging engines. The dredge quartered a crew of 21.

The dipper dredge *Col. M. B. Adams* was erected in 1912 at Wheeling, West Virginia, for the U.S. Engineer Department. The 400-ton dredge, shown loading scows on the Ohio River, could cut a swath up to 26-feet deep and 35-feet wide. An officer and eight men were quartered on the dredge.

This cutterhead suction dredge, owned by P. T. McCourt of Akron, Ohio, is seen working on the Cuyahoga River in Brooklyn, just south of Cleveland. The dredge's 16-inch pump was rated at 225 cubic yards of solid material per hour. The pump was powered by a 3-phase, 60-cycle, 2,200-volt General Electric motor. A 75-horsepower motor powered the cutter, and a 20-horsepower motor drove the 5-drum winch that operated the swinging lines, cutter head hoist, and spuds. Marion built five cutterhead suction dredges between 1914 and 1916.

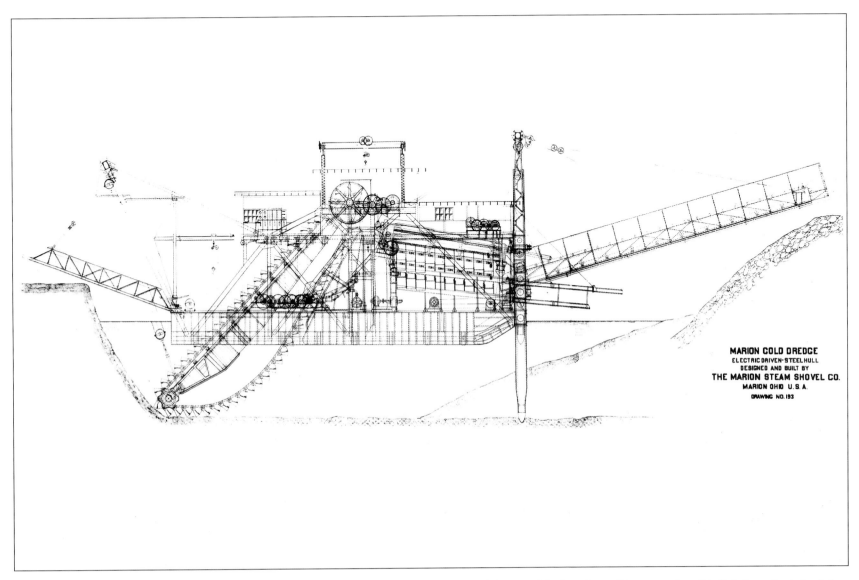

MARION GOLD DREDGE
ELECTRIC DRIVEN-STEEL HULL
DESIGNED AND BUILT BY
THE MARION STEAM SHOVEL CO.
MARION OHIO U.S.A.
DRAWING NO. 193

Design drawing for an electric-powered, steel hull, bucket-ladder gold dredge. Marion built 24 bucket-ladder dredges between 1904 and 1916 for prices between $48,000 and $178,000. Half of these dredges were erected in California, six in Canada's Yukon Territory, and five in Montana.

This bucket-ladder dredge is shown working a gold placer deposit near Oroville, California. The ladder of the dredge was 78 feet long and sported 64 buckets of 5 1/2-cubic foot capacities. She could dig between 80,000 and 100,000 cubic yards per month. The discharge conveyor at the stern measured 90 feet long.

Gold dredge No. 4 is shown in this photograph dated August 6, 1912, on Big Bear Creek in the Klondike River Valley of Canada's Yukon Territory. The dredge, built in 1912, was one of four built by Marion for the Canadian Klondyke Mining Company's placer operations. She was equipped with 15-cubic foot buckets, which were later built up to 16 cubic feet. With a displacement weight of over 3,000 tons, she is the largest wooden hull, bucket-ladder dredge built in North America. She was rebuilt several times and was finally retired in 1959. In 1992, she was restored by the Canadian government, and is part of the Dawson City-Klondike Historical Sites.

Introduced in 1972, the Marion Model M-4 is a rotary drill designed to bore holes 12 1/4 inches in diameter into solid rock for blasting. Marion also offered smaller M-1, M-2 and M-3 drills, a Mark I tandem drill with two nine-inch bits and the massive Model M-5; the M-5 wielded a 15-inch bit and could exert up to 120,000 pounds of pulldown, or down pressure, in drilling.

A Model 3560 hydraulic shovel is shown working for the Island Creek Coal Company in 1981. Powered by two Caterpillar or Cummins engines, the shovel boasted a total of 1,400 gross horsepower. The machine weighed 300 tons and had a 22-yard bucket. The Marion-Dresser logo above the cab windows is indicative of Dresser Industries' ownership of Marion from 1977 to 1992.

The 3560 was also offered as a hydraulic backhoe, with a standard 18-yard bucket or an optional 27-yard coal bucket. Total weight of the 3560 backhoe was 328 tons. Offered between 1981 and 1989, only eight 3560 shovels and backhoes were built. In contrast to the heavy commitment made by most other shovel and crane companies to hydraulic excavators and cranes in the 1980s, the 3560 was the only hydraulic machine Marion ever offered. The only other hydraulic excavator with which Marion was involved was a small telescoping-boom excavator produced briefly by subsidiary Quickway.

In 1973, Marion acquired rights to the Peerless Manufacturing Company's V-Con (Vehicle Constructors) Division, which had been formed four years previously with the intention of building the world's largest mine dump truck. The distinctive V-Con 3006 off-road dump truck, completed in 1971 with a capacity of 260 tons, fulfilled that goal. It weighed 175 tons, traveled on four pairs of straddle-mounted tires and was powered by a 12-cylinder, 3,000 horsepower Alco 251 locomotive diesel engine mounted under the dump bed. The engine powered a generator that supplied electricity to six motors, one in each of the four rear wheels and one in each of the outside front wheels. The 3006 was first tested at the Pima Copper Mine in Arizona, where it is seen here being loaded by a 191-M electric shovel. While several models were proposed and heavily marketed by Marion, no more V-Con trucks were built.

Initially conceived by Peerless Manufacturing, the 150-ton, 1,500 gross horsepower V-Con V-220 wheel dozer was introduced in 1975. Power was provided by a Detroit Diesel 16V-149T diesel generator that powered an electric motor in each wheel, mirroring the original wheel dozer propulsion designs of R. G. LeTourneau some thirty years previously. In spite of extensive demonstrations to the mine reclamation market in the American Midwest, only one other V-Con dozer, with a much different cab design, was built before Marion's Vehicle Constructors Division was disbanded.

The last machine in the book is, fittingly, the last new model developed by Marion before Bucyrus International acquired the company in 1997 to end the two companies' 113-year competition. Introduced in 1995, the 351-M weighs 1,300 tons and wields a hefty 57-yard bucket. Its predecessor, the 301-M, was introduced in 1985; the 301-M had the same size bucket but weighed 150 tons less. Just three bucketfuls from these enormous shovels will fill a 240-ton hauler to capacity. Bucyrus still offers this machine today as their Model 595B. *Photo courtesy of Eric Orlemann*

INDEX by Machine Type

MORE TITLES FROM ICONOGRAFIX

*THIS PRODUCT IS SOLD UNDER LICENSE FROM MACK TRUCKS, INC. MACK IS A REGISTERED TRADEMARK OF MACK TRUCKS, INC. ALL RIGHTS RESERVED.

All Iconografix books are available from direct mail specialty book dealers and bookstores worldwide, or can be ordered from the publisher.
For book trade and distribution information or to add your name to our mailing list and receive a **FREE CATALOG** contact:

Iconografix, PO Box 446, Dept BK, Hudson, Wisconsin, 54016   Telephone: (715) 381-9755, (800) 289-3504 (USA), Fax: (715) 381-9756

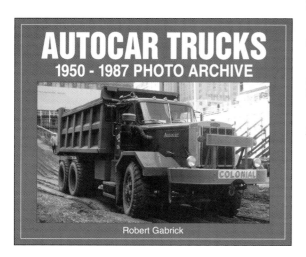

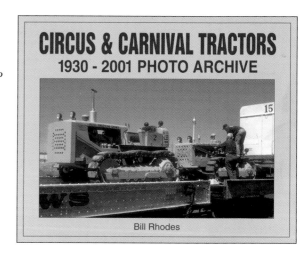
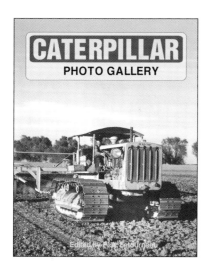
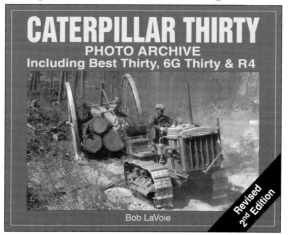
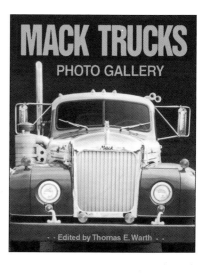